iPhone 16 Made

*A Step-by-Step Manual to Master the iPhone 16, Pro &
Pro Max with iOS 18 — For Seniors, Beginners &
Anyone Switching from Android or Older iPhones*

Georgette Howard

1

Table of Contents

Disclaimer

This guide is an independent publication and is not affiliated with, endorsed by, or sponsored by Apple Inc. "iPhone," "iOS," "Face ID," and all related marks are trademarks of Apple Inc., registered in the U.S. and other countries. All information in this book is provided for educational and informational purposes only. While every effort has been made to ensure accuracy, the author and publisher assume no responsibility for any errors, omissions, or outcomes resulting from the use of this guide. Always refer to Apple's official documentation or customer support for the most up-to-date and accurate information.

Preface

iPhone 16 Made Easy is the essential, step-by-step user guide for beginners and seniors who want to master the iPhone 16, Pro, and Pro Max—without confusion. Whether you're new to smartphones or upgrading from an older iPhone, this book makes iOS 18 simple, clear, and fun to learn.

Discover how to use your iPhone for calling, texting, photos, FaceTime, apps, privacy, Siri voice commands, iCloud backup, battery optimization, and more. With real-life examples, full-color image prompts, and helpful tips for all ages, this is your shortcut to becoming confident with your iPhone—fast.

Perfect for seniors, visual learners, and new iPhone users, this guide includes iPhone camera tricks, safety settings, and 30-day practice challenges to boost your confidence and creativity.

If you're looking for a complete iPhone 16 user guide for beginners, iPhone for seniors manual, or iOS 18 how-to book, this is the one to grab.

INTRODUCTION

Why This Guide Will Transform Your iPhone Experience

Welcome to your new iPhone adventure—whether you've just unboxed the sleek iPhone 16 for the first time, or you've been using iPhones for years but want to unlock its full power. This isn't just another tech manual filled with jargon or generic tips. *iPhone 16 Made Easy* was written with a real-world user in mind: someone who wants clarity, simplicity, and real value—step by step.

This guide is more than instructions. It's a companion that meets you at your current level and walks beside you as you grow from a casual tapper to a confident explorer—helping you use your device *smarter*, *faster*, and with more joy.

In a world of constant updates and ever-changing screens,

technology can sometimes feel overwhelming. But here's the truth: it doesn't have to be. With the right explanations, a few clever tricks, and the freedom to learn at your own pace, the iPhone 16 becomes less of a device and more of a tool for creativity, connection, and everyday ease.

By the end of this guide, you won't just know how to use your iPhone—you'll own it, inside and out.

Who This Book Is For

This guide was written for **three types of people**—each one as important as the next:

1. **Beginners & Seniors** – If this is your first iPhone, or if navigating new tech feels stressful, this book will walk you through each step slowly and clearly. No rushing, no assumptions—just patient, plain-language help.

2. **Everyday Enthusiasts** – If you've used iPhones before but haven't yet explored all the newest features in iOS 18 or iPhone 16's advanced camera and settings, you'll discover how much more your device can do with a little guidance.

3. **Power Users in Progress** – Even if you're somewhat advanced, this guide includes hidden features, automation tricks, and photography insights that go beyond the basics. You'll learn how to optimize, customize, and truly master your iPhone experience.

Whether you're 16 or 76, this book will speak your language and make even the most technical tools feel natural.

How to Use This Book: Read,

Practice, Master

Each chapter is carefully structured to *build your skills* as you go. Here's how to make the most of it:

- **Read** the instructions at your own pace. Every chapter is filled with real-life examples, screenshots (if included in your edition), and plain-English explanations.

- **Practice** the steps as you read. Try the tips in real-time on your iPhone to reinforce the learning.

- **Master** your iPhone by returning to the guide whenever you need a refresher. Use the index or chapter summaries to jump to what you need fast.

You don't have to read this book in one sitting. Think of it as your iPhone learning companion—there when you need it, waiting patiently when you don't.

Quick Overview of iPhone 16

Models: What's New, What's Different

Before diving in, it's helpful to understand which model you're using because not all iPhones are created equal.

Here's a simple breakdown of the current iPhone 16 lineup:

🖩 iPhone 16 (Standard)

- Perfect for everyday users.
- Smooth performance with Apple's A18 chip.
- Excellent camera system, Face ID, Dynamic Island, and OLED display.

📷 iPhone 16 Pro

- Designed for power users and creators.

- Includes ProMotion display (120Hz), advanced triple-lens camera, ProRAW and ProRes support, and titanium build.
- Ideal for photography, videography, and performance multitasking.

🎥 iPhone 16 Pro Max

- The powerhouse of the family.
- Largest screen size, highest battery life, and all Pro features plus additional zoom capabilities.
- Perfect for creatives, professionals, or those who simply want the best of the best.

All models run iOS 18, Apple's newest operating system filled with new customization features, smarter Siri interactions, and improved privacy tools.

Throughout this book, we'll flag where certain features are exclusive to the Pro and Pro Max models, so you can always find what applies to you.

Now that you've got your iPhone in hand and this guide in front of you—let's begin unlocking its full potential.

Welcome to the world of *iPhone 16 Made Easy*.

Let's get started.

Chapter 1

Unboxing the Future – Getting Started with iPhone 16

There's something unmistakably exciting about holding a brand-new iPhone in your hands. The seamless edges. The smooth glass. The polished titanium. But once the initial awe fades, a more practical question emerges:

"Now what?"

That's exactly what this chapter is here to answer.

Whether you're switching from an older model, transitioning from Android, or holding an iPhone for the very first time, this chapter will guide you through every step from choosing the right model to setting it up like a pro. No tech jargon, no confusion—just crystal-clear instructions and real-world help.

Choosing the Right Model: iPhone 16 vs. Pro vs. Pro Max

With three stunning models released in the iPhone 16 lineup, it's helpful to understand what sets them apart so you know exactly what features apply to your device.

iPhone 16 (Standard Model)

- **Display:** 6.1" OLED screen
- **Camera:** Dual-lens system with advanced Portrait and Night Mode
- **Chip:** Apple A18
- **Ideal For:** Casual users, students, parents, seniors

iPhone 16 Pro

- **Display:** 6.1" ProMotion display (120Hz refresh rate)
- **Camera:** Triple-lens system with macro, night, and ProRAW shooting

- **Chip:** A18 Pro for higher-speed tasks

- **Ideal For:** Mobile photographers, creatives, professionals

iPhone 16 Pro Max

- **Display:** 6.7" display—largest screen in the series

- **Camera:** Everything in the Pro, plus advanced zoom and sensor-shift stabilization

- **Battery:** Longest battery life in iPhone history

- **Ideal For:** Videographers, creators, power users

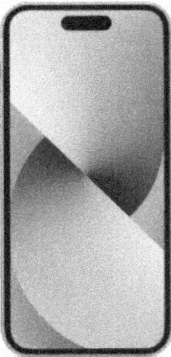  

iPhone 16   iPhone 16 Pro   iPhone 16 Pro Max

iPhone 16 lineup—choose the model that fits your lifestyle and needs.

What's in the Box: What Apple Includes (and What They Don't)

Apple continues its minimalistic packaging trend. While you'll get the essentials to start, you might be surprised by what's not included.

Included in the Box:

✓iPhone 16 / Pro / Pro Max

✓USB-C to USB-C charging cable

✓SIM card ejector tool (in regions with physical SIM support)

✓Apple logo sticker and basic manual

Not Included:

✗ Charging brick (power adapter)

✗ EarPods or headphones

✗ Case or screen protector

30

What's in the box—just the essentials. A charging adapter must be purchased separately.

First-Time Setup Made Simple

(Step-by-Step Activation)

Let's walk you through setting up your iPhone—calmly, clearly, and step-by-step.

Step 1: Power It On

- Press and hold the **Side button** until the Apple logo appears.

- You'll see the **Hello** screen in multiple languages.

Say hello to your new device—this is where your setup begins.

Step 2: Choose Language & Country

- Swipe up to begin.

- Select your preferred language and region.

Step 3: Connect to Wi-Fi

- Choose your home Wi-Fi network.

- Enter your password to connect.

Step 4: Set Up Face ID or Touch ID

- Position your face within the frame and rotate slowly.

- Follow the prompts to complete the scan.

Use Face ID to unlock, authenticate, and secure your phone effortlessly.

Step 5: Set a Passcode

- Choose a 6-digit passcode or create a custom one.

- Avoid simple patterns like "123456" or birthdates.

Step 6: Sign In or Create an Apple ID

- If you already have one, enter your Apple ID and password.

- Otherwise, tap "Forgot password or don't have an Apple ID?" to create a new one.

Apple ID, Passcode, and Face ID Setup

Your Apple ID is your digital key to the Apple ecosystem. It links your phone to iCloud, App Store, Messages, and more.

Apple ID Tips:

- Use a secure, regularly accessed email address.

- Turn on **two-factor authentication** for security.

- Don't share your credentials—ever.

Caption: Your Apple ID is your gateway to apps, cloud storage, and security.

Connecting to Wi-Fi, iCloud, Bluetooth & Cellular Networks

Let's make sure you're connected to everything your iPhone needs.

1. Wi-Fi Setup

Go to: **Settings > Wi-Fi**

- Select your network

- Tap "Auto-Join" for convenience

- Toggle off when using public Wi-Fi for privacy

2. iCloud Configuration

Go to: **Settings > [Your Name] > iCloud**

- Toggle ON what you want to sync: Photos, Contacts, Messages, Notes

- Ensure Find My iPhone is turned ON

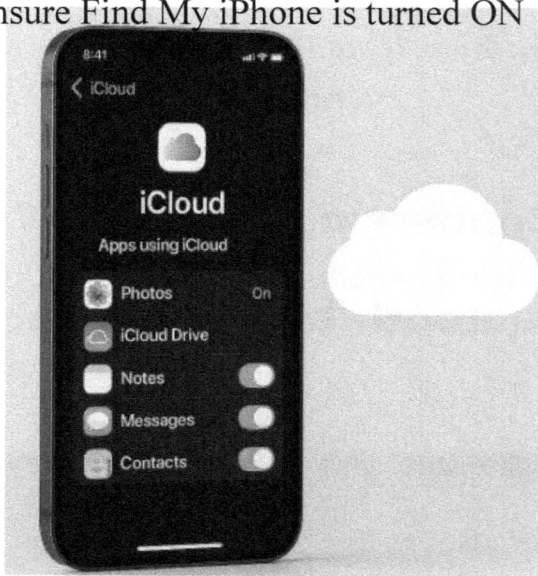

iCloud keeps your data safe, backed up, and synced across devices.

3. Bluetooth Setup

Go to: **Settings > Bluetooth**

- Turn it ON

- Pair AirPods, speakers, or accessories as needed

4. Cellular Activation (SIM or eSIM)

- Insert physical SIM using the ejector tool, or

- Follow on-screen steps for eSIM activation using your carrier's QR code

Updating to the Latest iOS 18 Version

Apple's iOS 18 introduces new widgets, lock screen customizations, smarter Siri, and more.

How to Update:

Go to: **Settings** > **General** > **Software Update**

- Tap **Download and Install**

- Connect to Wi-Fi and plug in your phone if battery

is low

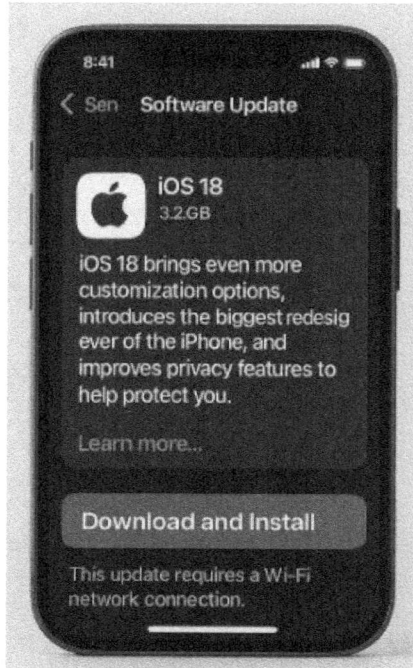

Keep your iPhone up to date with the latest features and security improvements.

Recap

- ✓ Choose the right model
- ✓ Unbox essentials

✓ Activate and connect

✓ Secure with Apple ID and Face ID

✓ Sync to Wi-Fi, iCloud, Bluetooth

✓ Update to iOS 18

With the basics now complete, your iPhone 16 is ready to serve as your daily assistant, creative tool, and personal connection hub. In the next chapter, we'll dive deep into the **iOS 18 interface** and teach you how to confidently navigate the Home screen, widgets, control center, app switching, and more.

Chapter 2

Once your iPhone is up and running, the real fun begins. iOS 18 brings a sleek, intuitive design to the screen you'll use every day but if you're new to the ecosystem, it can feel like entering a city without a map.

Don't worry, we'll make it feel like home.

This chapter will walk you through the **Home Screen**, **Control Center**, **Widgets**, **Dynamic Island**, and more— with clear guidance and visual cues so you can navigate your iPhone with complete confidence.

Home Screen Layout & Control Center Explained

Your Home Screen is the central hub of your iPhone. It's

where your most-used apps live and where you'll return whenever you exit an app.

Key Elements:

- **Dock (bottom row)** – Your favorite apps that stay visible on all Home Screens.

- **App icons** – Tap to open, press-and-hold to rearrange or delete.

- **Pages** – Swipe left or right to move between app pages.

The Home Screen—your daily starting point. Customize with apps, widgets, and folders.

Control Center: Your Quick Command Panel

Swipe down from the **top-right** corner to open the Control Center. Here, you'll find:

- Wi-Fi, Bluetooth, Airplane Mode
- Flashlight, Screen Brightness
- Volume Control, Music Player
- Focus Modes, Screen Mirroring

Caption: Swipe from the top-right to access the Control

App Library vs. Folders

As you install apps, your screen can get crowded but Apple offers two powerful ways to stay organized.

Folders

- Drag one app icon over another to create a folder.

- Name it something intuitive like "Finance" or "Games."

- Tap to open, press-and-hold to rename or remove apps.

App Library

- Swipe **all the way left** on your Home Screen.

- You'll find your apps auto-grouped by category (Productivity, Social, etc.).

- Use the search bar at the top to find any app fast.

Use folders for custom grouping; the App Library handles the rest automatically.

Using the Dynamic Island & Live Activities

If you're using iPhone 16 or 16 Pro, you've probably noticed a pill-shaped space at the top of your screen. That's **Dynamic Island**—a live info center that keeps you updated without switching apps.

What It Shows:

- Incoming calls

- Music playing

- AirDrop transfers

- Timer countdowns

- Face ID recognition

- Live Activities like food delivery or Uber

Tap it to expand and interact, or just glance for live updates.

Dynamic Island keeps live info visible while you multitask with ease.

iOS 18 Widgets & Lock Screen Customization

Widgets are miniature versions of apps that display real-time info without needing to open anything.

How to Add a Widget:

1. Tap and hold an empty part of the Home Screen.
2. Tap the + icon in the top-left.
3. Browse or search for the widget you want.
4. Tap **Add Widget** to place it.

Lock Screen Customization (iOS 18 Update)

- Press and hold your Lock Screen.
- Tap Customize to change wallpaper, font, and widget layout.

- You can even assign Lock Screens to different Focus Modes.

Customize your Lock Screen and Home Screen with interactive, helpful widgets.

Notification Center & Focus Modes

Stay informed—without feeling overwhelmed.

Notification Center

- Swipe down from the **top-middle** of your screen.
- See recent app alerts, messages, and missed calls.

Focus Modes (Work, Sleep, Personal, etc.)

- Go to: **Settings > Focus**

- Set up which apps or people can notify you during that mode

- Automate it by time, location, or app usage

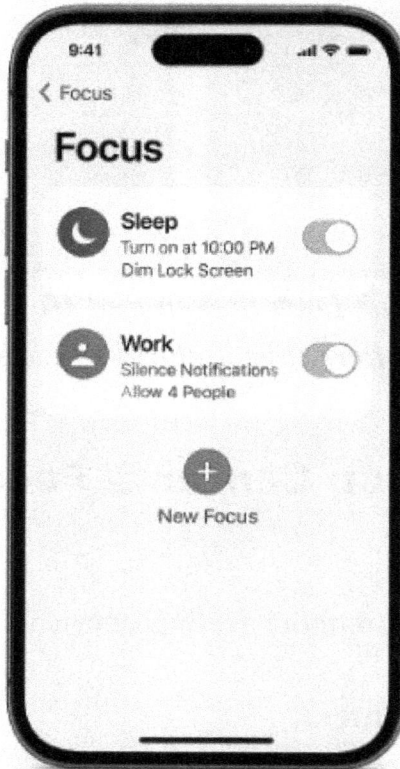

Focus Modes help you stay in control of your notifications throughout the day.

Beginner Tips: Swiping, Searching, and Multitasking

Master these gestures and you'll feel like a pro in no time:

- **Swipe Up** – Go Home (from any app)

- **Swipe Down (middle)** – Open Notification Center

- **Swipe Down (left or right)** – Search apps, contacts, or web

- **Swipe & Hold** – View all open apps (App Switcher)

- **Swipe Between Apps** – On iPhone 16 Pro/Max, swipe the bottom edge to switch between apps quickly

iOS 18 Swiping Gestures

Upward swipe

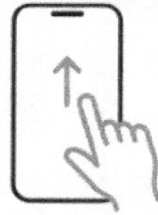

Top edge swipe

Right edge swipe

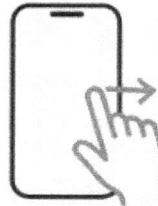

Left edge swipe

iOS 18 gestures help you navigate with your fingertips—no buttons required.

Accessibility Features for Seniors & Visual Assistance

Apple has built iOS 18 with inclusivity in mind. These tools are a game-changer for seniors or anyone needing visual or physical support.

Top Features to Turn On:

- **VoiceOver** – Reads out what's on your screen

- **Zoom** – Magnify part of the screen with gestures

- **Magnifier** – Use the camera as a digital magnifying glass

- **Display Accommodations** – Invert colors, reduce white point

- **Spoken Content** – Have Siri read emails, books, and more

- **Touch Accommodations** – Adjust screen sensitivity

- **Hearing Devices** – Pair hearing aids via Bluetooth

- **AssistiveTouch** – On-screen virtual buttons for easier navigation

Go to: **Settings > Accessibility** to explore.

Make your iPhone more comfortable with easy-to-use accessibility tools.

Recap & Practice

In this chapter, you learned how to:

- ✓ Navigate your Home Screen and Control Center

- ✓ Organize your apps with folders and App Library

- ✓ Use Dynamic Island and Live Activities

- ✓ Add widgets and customize your Lock Screen

✓ Manage notifications and Focus Modes

✓ Master essential gestures

✓ Enable accessibility tools for better usability

Every swipe, tap, and hold brings you closer to mastery.

Chapter 3

Core Essentials – Mastering Everyday Apps & Functions

Your iPhone 16 isn't just a phone, it's your digital Swiss Army knife. Whether you're calling a loved one, checking tomorrow's forecast, browsing the web, or setting a reminder to water your plants, your iPhone is ready to help. In this chapter, we'll walk through the essential Apple apps you'll use every day and teach you how to get the most out of each one.

These apps are built into iOS 18 and require no downloads. All you need is curiosity and a few minutes of practice.

Phone, Contacts, Messages & FaceTime

These four apps are the foundation of your communication world.

Phone

- Open the **Phone** app to make or receive calls.

- Use the **Keypad** to dial a number manually.

- Tap **Recents** to see missed or recent calls.

- Go to **Favorites** to save people you call frequently.

- Use **Voicemail** to listen to recorded messages.

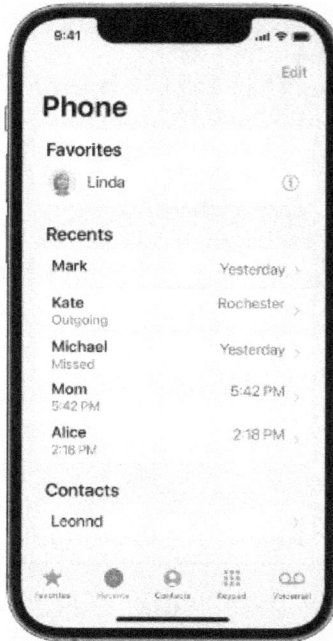

The Phone app keeps all your calls, contacts, and voicemail organized and easy to access.

Contacts

- Open the **Contacts** app or go to **Phone > Contacts.**

- Tap the + to add a new contact with phone number, email, and photo.

- You can add custom labels like "Doctor," "Emergency," or "Family."

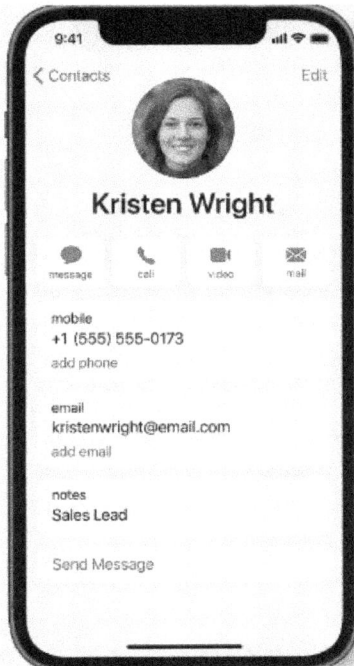

Create detailed contacts with phone numbers, birthdays, and notes.

Messages

- Use **Messages** for SMS (green) and iMessage (blue).

- Tap the **pencil icon** in the top right to start a new message.

- Use stickers, images, GIFs, and voice messages.

- Long-press a message to **react** or **reply inline**.

57

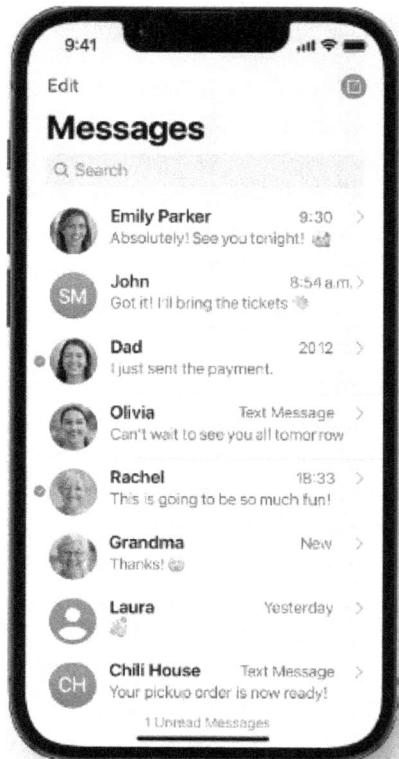

Express more than words—send emojis, stickers, or quick reactions in iMessage.

FaceTime

- Open FaceTime to make a video or audio call.

- Tap **New FaceTime**, choose a contact, and hit the green **FaceTime** button.

- Add effects, use Portrait Mode, and share your screen.

Connect face-to-face in real time with built-in FaceTime, even across Apple devices.

Safari: Smart Browsing & Private Tabs

Safari is Apple's secure and fast web browser.

How to Use:

- Open the **Safari app** and type in a website or search term.

- Tap the **Tab icon** in the bottom right to manage multiple websites.

- Tap **Private** to enter incognito browsing (no history stored).

- Use **Reader Mode** (Aa button) for distraction-free reading.

Browse smart and safe with Safari's Private Tabs and Reader Mode.

Mail: Add Email Accounts, Filters & Safe Practices

Set up Mail to send and receive all your email in one place.

Setup:

- Go to **Settings > Mail > Accounts > Add Account**.

- Choose iCloud, Gmail, Outlook, or add manually.

- Toggle on **Mail, Contacts, and Calendars**.

Smart Tips:

- Swipe left on emails to delete, archive, or flag.

- Use **VIP** to mark important contacts.

- Avoid tapping unknown links—report spam or phishing.

Keep your inbox under control with folders, filters, and quick swipe gestures.

Calendar & Reminders for Daily Organization

Calendar

- Use Calendar to add appointments, birthdays, and events.

- Tap the + button to create a new entry.

- Add location, alert times, and invite others.

Reminders

- Create grocery lists, to-dos, or daily goals.

- Use **Smart Lists** like "Scheduled" and "Flagged."

- Add **location-based alerts** (e.g., "Remind me to buy eggs when I arrive at the store").

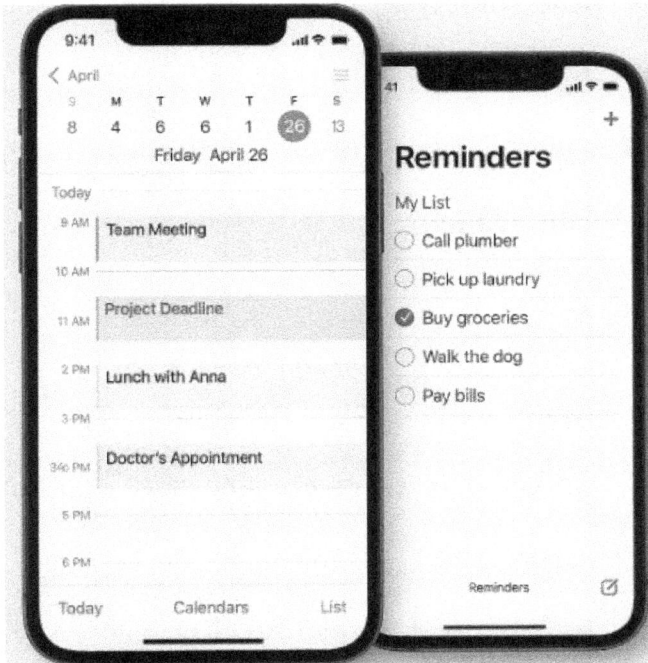

Stay organized with color-coded calendars and checkable to-do lists.

63

Notes, Voice Memos, and Quick Actions

Notes

- Open **Notes** for quick writing, checklists, and sketches.

- Use folders to group ideas (e.g., Recipes, Travel, Ideas).

- Scan documents or handwritten notes using the **camera icon.**

Voice Memos

- Record voice notes, interviews, or ideas on the go.

- Tap the **red circle** to start, pause, and save recordings.

- Rename and trim audio clips easily.

Turn your iPhone into a pocket notebook and recorder.

Apple Maps, Compass, and

Weather: Smart Navigation

Maps

- Use Maps for real-time directions, public transit, and walking routes.

- Tap **Go** after entering a destination.

- Use Look Around (like Google Street View) for visual previews.

Compass

- Open Compass for direction and elevation data.
- Great for hiking, stargazing, or basic orientation.

Weather

- Check temperature, hourly forecast, and 10-day trend.
- Get alerts for severe weather, air quality, and rainfall.

Navigate your world confidently with built-in Maps, Compass, and Weather apps.

Beginner Pro Tip: Using Siri for

Everyday Tasks

Siri is your voice-activated assistant—and one of the most underused tools.

Try Saying:

- "Call Sarah."

- "Remind me to take my medicine at 8 PM."

- "Set a timer for 10 minutes."

- "What's the weather tomorrow?"

- "Open Camera."

Let Siri handle your reminders, calls, searches, and more—just by asking.

Recap: Mastering the Core

You now know how to:

- Make calls, send texts, and video chat

- Browse the web with privacy and speed

- Manage your email and calendar

- Create notes, checklists, and voice memos

- Navigate your world with smart Apple apps

- Let Siri do the heavy lifting for everyday tasks

These are the everyday tools that make your iPhone a reliable, personal assistant. Master these, and you're well on your way to full iPhone confidence.

Chapter 4

Camera Confidence – From First Snap to Photo Genius

Your iPhone 16 is more than a phone, it's one of the most powerful pocket cameras ever made.

Whether you're capturing your grandchild's smile, the glow of a sunset, or your dog mid-air, the iPhone 16 camera is built to make anyone look like a pro. In this chapter, you'll discover how to unlock its full potential with simple techniques and powerful features.

Let's start by getting familiar with the Camera app—your new best friend for memories.

Mastering the iPhone 16 Camera

Interface (Step-by-Step)

When you open the Camera app, it may look simple—but it's packed with powerful tools.

Interface Overview:

- **Shutter Button:** Large white circle (tap once for photo)

- **Mode Selector:** Swipe left/right (Photo, Video, Portrait, etc.)

- **Zoom Buttons:** 0.5x, 1x, 2x, 3x (varies by model)

- **Flash:** Lightning bolt icon (Auto, On, Off)

- **Live Photo:** Yellow circle (for motion + sound)

- **Settings Panel:** Swipe up or tap small arrows ↑

- **Filters & Exposure:** Swipe up to access controls like timer, filters, and aspect ratio

What Each Lens Does (Wide, Ultra-Wide, Telephoto on Pro Models)

Your iPhone 16 may have 2 or 3 lenses depending on your model:

iPhone 16 (Standard):

- **Main Wide Lens (1x):** Great for everyday shots
- **Ultra-Wide Lens (0.5x):** Fits more into the frame—perfect for landscapes, rooms, groups

iPhone 16 Pro / Pro Max:

Wide Lens (1x): Standard, high-quality shots

Ultra-Wide (0.5x): Dramatic and wide views

Telephoto (2x–5x): Close-up shots from a distance without losing clarity

| Ultra Wide (0.5x) | Wide (1x) | Telephoto (3x) |

Each lens has a different purpose. Use wide for general shots, ultra-wide for scenery, and telephoto for detail.

Portrait Mode, Night Mode, Macro Mode

Portrait Mode

- Creates a depth-of-field blur behind the subject

- Best for people, pets, and still-life

- Adjust blur level after shooting via **Edit > f/stop**

Night Mode

- Automatically activates in low-light

- Uses extended exposure (1–5 seconds)

- Keep the phone steady—use a table or tripod for best results

Capture stunning low-light photos with Night Mode— no flash needed.

Macro Mode (Pro Models Only)

- Activated automatically when shooting very close

- Perfect for flowers, textures, food, or jewelry

- Get within 2 inches of your subject

Live Photos vs. Still Shots: When & How to Use Them

Live Photos:

- Captures 1.5 seconds before and after your shot

- Includes motion + sound

- Great for candid, emotional, or action moments

Toggle it ON/OFF with the **Live circle icon** at the top of your screen.

Pro Tip: You can turn a Live Photo into a still, loop, bounce, or long exposure.

Live Photos add motion and life to your shots—great for memories that move.

Burst Mode, Action Shots, and Motion Focus

Burst Mode:

- Hold down the **Shutter button** in Photo mode

- Your iPhone takes multiple rapid shots

- Perfect for fast-moving subjects (kids, pets, sports)

Find the best frame afterward by tapping **Photos > Select** in the burst set.

Action Shots:

- Use Burst Mode or Video Screenshots
- Track your subject in frame
- Bright lighting helps freeze motion

Motion Focus (Tracking Autofocus):

- Tap your subject to lock focus
- Follow them as they move—focus stays sharp

Burst Mode helps you capture the perfect frame—
especially when timing matters.

iPhone Video Features: Cinematic Mode, 4K Recording, Slow-Mo

Cinematic Mode

- Adds depth-of-field (blurry background) to video

- Automatically shifts focus between subjects

- Editable after recording

4K & HD Recording

- Open **Settings > Camera > Record Video**

- Choose between 720p, 1080p, or 4K at 24/30/60 fps

Slow-Mo

- Great for sports, dancing, or splashing water

- Recorded in 120 or 240 fps

- Trim and control slow-motion effect in editing

Record movie-like videos with stunning focus effects using Cinematic Mode.

Practical Guide: Taking Perfect Photos of People, Nature, Food, and Pets

People

- Use **Portrait Mode**
- Tap face to set focus
- Use natural light near windows
- Avoid zooming—move closer

Nature

Use **Ultra-Wide Lens**

Add foreground elements (tree branches, flowers) for depth

Try **Panorama Mode** for wide landscapes

Food

- Use **Top-Down** angle

- Clean background

- Use natural lighting

- Try Macro for texture

Pets

- Get to eye level

- Use Burst Mode for movement

- Use treats or toys for attention

- Avoid flash

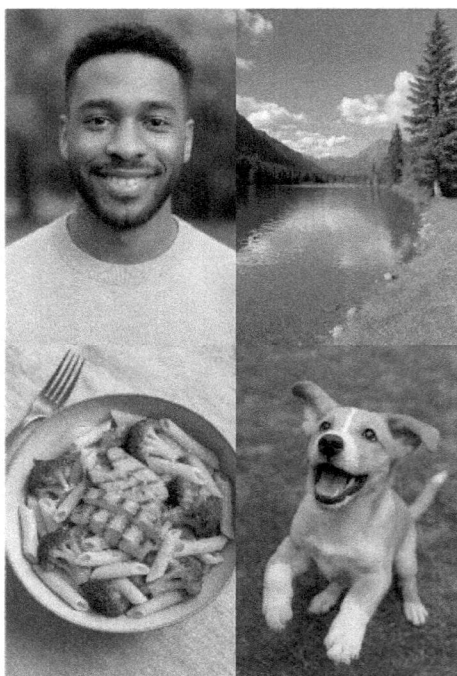

Every subject shines with the right lens, light, and angle.

Recap: From Tap to Mastery

In this chapter, you've learned how to:

- ✓ Navigate the Camera interface with confidence

- ✓ Choose the right lens for the right moment

- ✓ Master Portrait, Night, and Macro modes

- ✓ Capture the magic of movement with Burst and Live Photos

✓ Shoot cinematic-quality video

Take stunning shots of people, nature, food, and pets.

Chapter 5

Edit Like a Pro – Photos, Filters & Albums

So you've taken the perfect shot or maybe just a pretty good one. Now what?

This chapter is where your iPhone photos go from ordinary to unforgettable. You'll learn how to edit with precision, organize your photo library like a pro, and share your moments beautifully and securely. Best of all, you'll use tools that are already built into your iPhone—no extra apps required.

Using the Built-in Photos App for Editing

The Photos app is more than a storage gallery. It's a powerful editor designed to help you enhance your images in just a few taps.

How to Begin Editing:

1. Open **Photos**

2. Tap any photo

3. Tap **Edit** in the top-right corner

You'll now see a series of tools: **Auto-Enhance, Adjustments, Filters, Crop/Rotate,** and more.

The built-in editor in the Photos app gives you powerful tools to enhance any image instantly.

85

Crop, Adjust, Auto-Enhance, and Filters

Auto-Enhance

- Tap the magic wand icon to instantly improve lighting and color.

- Best for quick fixes without diving into details.

Crop & Rotate

- Tap the square-and-arrow icon.

- Pinch to zoom and drag the frame.

- Use rotate or flip tools for alignment.

- Straighten crooked shots with the slider.

Adjust

- Tap the dial icon for fine-tuned control:

  - Exposure – Brighten or darken

  - Brilliance – Midtone clarity

- Highlights/Shadows – Balance extremes

- Contrast, Brightness, Saturation, Warmth, Sharpness, etc.

Filters

- Tap the three-circle icon.

- Swipe through built-in filters like **Vivid, Dramatic, Silvertone, Noir**

- Adjust filter strength by dragging the slider

You don't need to be a pro—just a few tweaks can bring your photos to life.

Organizing Albums and Favorites

A cluttered photo library can make finding your best shots feel like searching for a needle in a haystack. Here's how to stay organized:

Favorites

- Tap the **heart icon** on any photo to mark it as a favorite.
- Access all favorites under **Albums > Favorites**.

Albums

- Open **Photos > Albums > +**
- Tap **New Album** and give it a name (e.g., "Vacation 2025" or "Family")
- Select the photos to add

Smart Albums

Apple automatically creates albums like **Selfies,**

Portraits, Screenshots, Videos, and more.

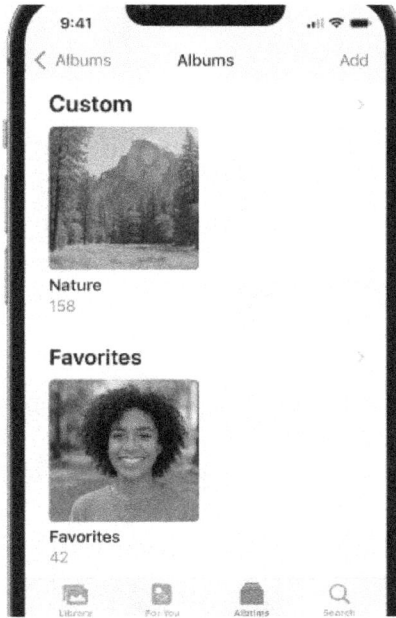

Albums and Favorites help you instantly access what matters most.

iCloud Photo Library: What It Is & How to Use It

iCloud Photos syncs your entire photo library across all your Apple devices—iPhone, iPad, Mac, and even the web.

To Turn It On:

- Go to **Settings > [Your Name] > iCloud > Photos**

- Toggle ON **Sync this iPhone**

Benefits:

- Automatically backs up your photos

- Access your images from any Apple device

- Saves storage using **Optimize iPhone Storage**

Turn on iCloud Photos to sync and back up your images seamlessly across devices.

Backing Up Your Photos Safely

Beyond iCloud, here are extra ways to make sure your precious memories are never lost.

Manual Backup to a Computer

- Use Finder on Mac or iTunes on PC
- Connect your iPhone via USB
- Select **Back Up Now**

External Drive Backup

- Plug in a **Lightning-to-USB** drive
- Use the Files app to copy your images

Google Photos / Dropbox / OneDrive

- Download a cloud app
- Enable automatic photo uploads
- Acts as a second layer of backup

Use a cloud backup and/or external drive for peace of mind and photo security.

Creating Photo Memories and Slideshows

iOS 18 automatically generates stunning Memories—short video slideshows complete with music and transitions.

How to View or Create:

- Open **Photos > For You**

- Scroll to Memories

- Tap **one to watch**

- Tap the ••• **More button** > **Add to Memories** to make your own

You can **edit the music, title, and photo order**, or export it as a video.

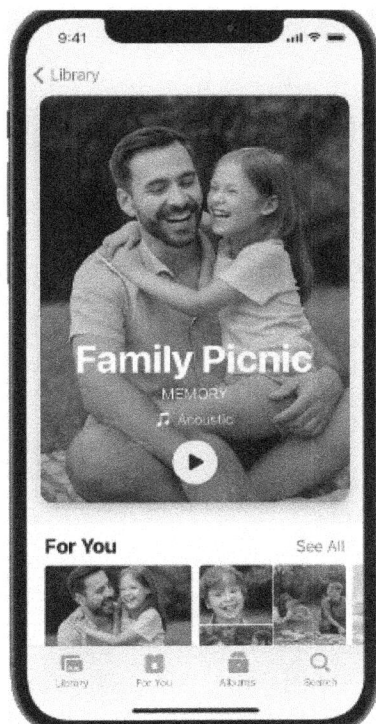

Let your iPhone tell your story through beautiful, auto-curated Memories.

Sharing Photos via AirDrop, Messages, and Shared Albums

AirDrop (Nearby Apple Devices)

- Open photo > Tap **Share icon > AirDrop**

- Select nearby user

- Make sure both Bluetooth and Wi-Fi are ON

Messages

- Open Messages > Tap camera or photo icon

- Select your image

- Add a message and tap send

Shared Albums

- Go to Photos > **Albums > + > New Shared Album**

- Invite people via iMessage or email

- Contributors can add photos and comments

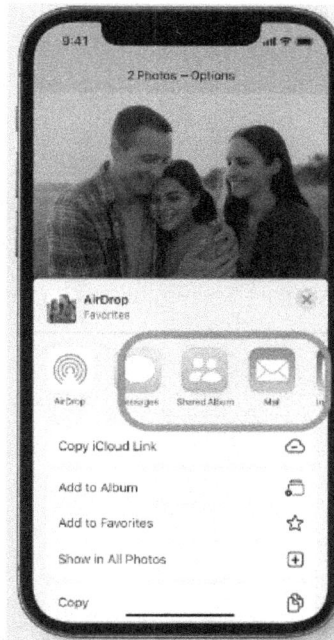

Whether near or far, share your moments instantly and beautifully.

Recap: Your Photo Power Toolkit

You now know how to:

✓ Edit your photos with professional tools

✓ Use Auto-Enhance, Crop, Filters, and Adjustments

✓ Create organized albums and mark favorites

✓ Use iCloud Photos to sync and protect your library

✓ Backup your media to cloud and local drives

✓ Craft meaningful Memories and slideshows

✓ Share easily via AirDrop, Messages, or Shared Albums

With just a few taps, your iPhone becomes your personal photo lab and your memories are preserved, enhanced, and ready to share.

Chapter 6

App Power – Best Apps for Productivity, Fun, and Security

Your iPhone becomes truly powerful when you start using the right apps. Apple's App Store offers over a million choices but the goal of this chapter is to make things simple, practical, and safe.

Whether you're a first-time iPhone user or a long-time Apple fan, these apps and the skills you'll learn here will help you do more with less effort.

Top Must-Have Apps for Beginners

These default Apple apps come pre-installed and are incredibly helpful when used to their full potential.

Health

- Tracks your steps, heart rate, and health records

- Syncs with Apple Watch and third-party fitness apps

- Set up **Medication Reminders, Sleep Goals**, and more

Translate

- Translates speech, text, or conversations in real-time

- Supports dozens of languages

- Great for travel or international communication

Calculator

- Use basic mode in portrait

- Rotate your iPhone to landscape **for Scientific Calculator**

- Access calculation history in **Control Center**

Files

- A simple file manager for PDFs, images, downloads, and more

- Connects to iCloud Drive, Google Drive, Dropbox, etc.

- Organize files with folders and tags

Apple's built-in apps are smarter than they look— explore them before searching the App Store.

Essential Free and Paid Apps for All

Users

Here are trusted, high-quality apps used by millions to boost productivity, creativity, and entertainment:

Productivity

- **Google Calendar** – Clean scheduling across devices
- **Evernote / Notion** – Organize notes, tasks, and projects
- **Todoist** – Powerful task manager with smart reminders

Creativity

- **Canva** – Easy design tool for flyers, social media, and more
- **Snapseed / Lightroom** – Professional-level photo editing

- **iMovie / CapCut** – Great for video editing on the go

Entertainment

- **Spotify / Apple Music** – Stream music and podcasts
- **Netflix / YouTube / Tubi** – Watch shows, movies, or DIY videos
- **Libby** – Borrow free eBooks and audiobooks from your library

Security & Privacy

- **1Password / LastPass** – Manage and autofill strong passwords
- **Authy** – Two-factor authentication for online accounts
- **ProtonMail / DuckDuckGo Browser** – Private email and browsing

Hidden iOS Apps Worth Exploring

iOS includes a few powerful apps that don't appear on the Home Screen by default but they're built in and ready to go.

Magnifier

- Turn your iPhone into a digital magnifying glass
- Great for reading menus or tiny text

Go to: **Settings > Accessibility > Magnifier > On**

Measure

- Use augmented reality to measure real-world objects
- Tap to mark start and end points
- Includes a level tool for checking if something is straight

Shortcuts

- Automate daily tasks with one tap

- Examples: "Text my spouse I'm on my way," or "Start my morning playlist"

- Find pre-made shortcuts in the Gallery tab

These "hidden gems" can make everyday tasks faster, easier, and smarter.

App Store Safety: Avoiding Scams,

Managing Subscriptions

Not all apps are created equal. Some hide fees or misuse your data. Here's how to stay safe and in control:

Check Before You Download

- Read reviews & ratings
- Look at developer info—avoid vague or foreign names
- Watch out for overly aggressive pricing in "Free Trials"

Manage Subscriptions

Go to: **Settings > Your Name > Subscriptions**

- See all active and expired app subscriptions
- Cancel any subscription with one tap

App Permissions

Go to: **Settings > Privacy & Security**

- Review access to Camera, Microphone, Contacts, Location, etc.

- Revoke access from apps that don't need it

Managing Screen Time & App Limits

Apple lets you monitor and limit your phone usage to help build healthy habits.

View Screen Time:

Go to: **Settings > Screen Time**

- View app usage, pickups, and notifications

- Set **App Limits** (e.g., 30 minutes daily for Instagram)

- Enable **Downtime** to restrict access during specific hours

Parental Controls (For Family Sharing)

- Limit apps and content for your child's device

- Block in-app purchases or explicit material

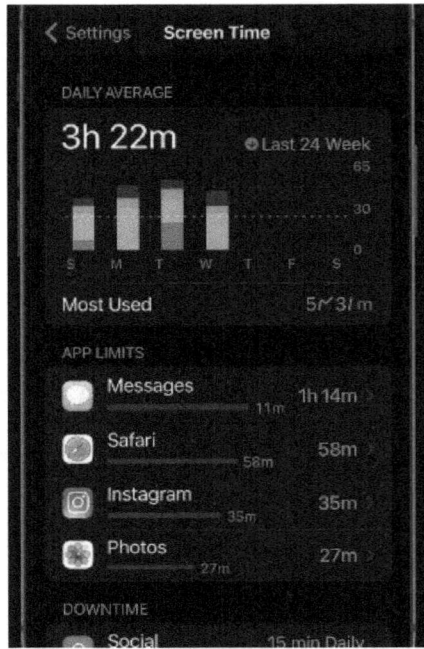

Balance productivity and well-being with Screen Time tracking and app restrictions.

How to Offload Unused Apps

Without Deleting Data

If you're running out of space but don't want to lose your data, **Offload** is the perfect solution.

How It Works:

- Removes the app from your iPhone

- Keeps its data and documents safe

- Reinstalls the app with one tap when needed

To Enable:

Go to: **Settings > iPhone Storage**

- Tap an app > Select **Offload App**

- Or enable automatic offloading at the top

Recap: Smart App Usage for a Smarter iPhone

By now, you've learned how to:

- ✓ Use Apple's most helpful built-in apps

- ✓ Discover and install essential third-party tools

- ✓ Explore hidden iOS apps for clever solutions

- ✓ Stay safe while navigating the App Store

- ✓ Limit distractions with Screen Time and app limits

- ✓ Offload unused apps without losing data

In the next chapter, we'll dive into Messages & Social Life—where you'll learn how to connect through texting, FaceTime, and more in style.

Chapter 7

Messages & Social Life – Mastering Communication

on iPhone

From heartfelt texts to joyful video calls, your iPhone 16 offers countless ways to connect. In this chapter, we'll explore how to make the most of iMessage, FaceTime, call settings, and even your favorite social apps—so you can stay close to those who matter most.

Mastering iMessage: GIFs, Stickers, Threads, Reactions

iMessage isn't just about texting, it's a vibrant space where you can express yourself in fun, interactive ways.

Basic Features

- Blue bubbles = iMessage (free, encrypted)

- Green bubbles = SMS (standard carrier messaging)

Add Media

- Tap the **camera** to take or send photos

- Tap the **App Store** icon to send stickers, emojis, or mini-apps

- Tap the #**images** icon to search and send GIFs

React to Messages

- Press and hold a message

- Choose from heart, thumbs up/down, haha, exclamation, or question mark

Threads & Replies

- In group chats, swipe right on a message to reply inline

- Keeps conversations organized

Bring your messages to life with reactions, GIFs, stickers, and reply threads.

Group Chats, Pinning, and Audio Messages

Group Chats

- Add multiple people by tapping the **compose icon** > **To**: field

- Name your group by tapping **details** > **Change Name and Photo**

- Mute or leave groups if needed via the **info icon**

Pin Conversations

- Swipe right on a conversation > Tap **Pin**

- Keeps frequent chats at the top of Messages

Send Audio Messages

- Tap and hold the **microphone icon** to record

- Release to send, swipe up to lock and continue recording

- Playbacks are temporary unless saved

FaceTime Tricks: SharePlay, Portrait Mode, Screen Sharing

FaceTime isn't just for video calling anymore, iOS 18 has made it smarter and more fun.

SharePlay

- Watch movies, listen to music, or play games together in real time

- Tap **FaceTime > Start New Call > Share Content**

Portrait Mode

- Blurs your background during calls

- Open **Control Center > Effects > Portrait** during a call

Screen Sharing

- Share your screen to walk someone through a task or show what you're doing

- Tap **Screen Share** icon during FaceTime

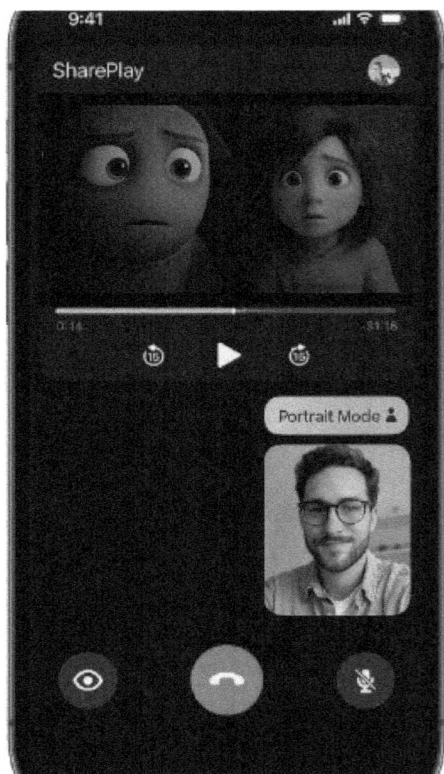

FaceTime now lets you share experiences, not just your face.

Managing Call Blocking, Spam Prevention, and Do Not Disturb

Not every call or message deserves your attention. Here's how to stay in control:

Block Unwanted Contacts

- Go to: **Phone > Recents > Info icon > Block this Caller**

Silence Unknown Callers

- Go to: **Settings > Phone > Silence Unknown Callers**
- Unknown numbers go straight to voicemail

Do Not Disturb / Focus

- Swipe to open **Control Center**
- Tap **Focus > Do Not Disturb**
- Customize to allow calls from Favorites only

Silence Junk Texts

- Go to: **Settings > Messages > Filter Unknown Senders**

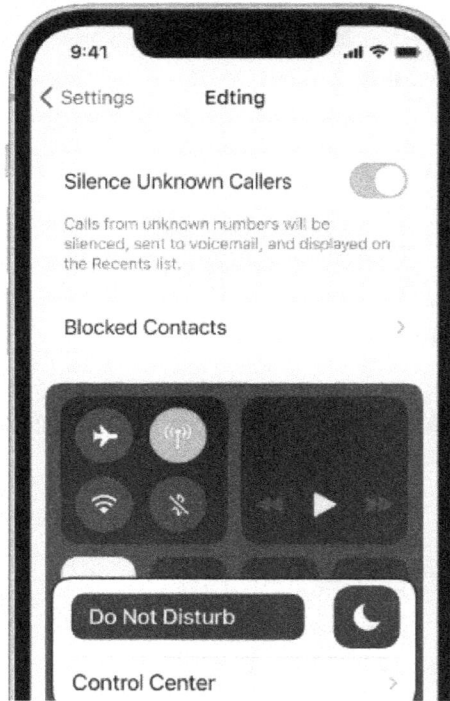

Protect your peace with smart call filtering and Do Not Disturb settings.

Social Media Integration with iPhone Features

The iPhone makes it easy to share moments across apps and control how they appear on your Home Screen.

Share Sheet

- Tap the **Share icon** from any photo, webpage, or note

- Choose to send via Messages, Instagram, Facebook, WhatsApp, or email

- Scroll to customize your Share Sheet with shortcuts

iOS Widgets for Social Apps

- Long press on the Home Screen > Tap +

- Add widgets for your favorite apps like Twitter (X), Instagram, or TikTok

- Get real-time updates or quick access to your social feed

Auto-Fill & Password Sync

- iCloud Keychain saves login info

- Face ID auto-fills usernames and passwords in apps or browsers

Share smarter and stay connected faster with integrated iPhone features.

Recap: Stay Connected with Confidence

You've now mastered how to:

✓ Use iMessage like a pro with GIFs, reactions, and threads

✓ Handle group chats, pin important conversations, and send audio messages

✓ Make the most of FaceTime with Portrait Mode, SharePlay, and screen sharing

118

✓ Block spam, silence unwanted calls, and use Focus Modes

✓ Integrate social media into your iPhone workflow

In the next chapter, we'll explore privacy and security features—so your information stays safe while you stay social.

Chapter 8

iOS 18 Power User Tips – Secrets Apple Doesn't Advertise

So you've mastered the basics but what if your iPhone could do even more?

Apple hides some of its most powerful features beneath the surface. This chapter is your roadmap to pro-level tools and automation secrets that most users never discover.

Hidden Gestures & Swipe Tricks

iOS 18 is gesture-driven. Once you learn a few hidden swipes, your phone becomes faster to use and more fluid to control.

Must-Know Gestures:

- **Swipe with three fingers** left/right → Undo/Redo typing or edits

- **Pinch with three fingers** → Copy or paste in Notes, Messages, Mail

- **Swipe right on the Home Bar** → Instantly switch between apps

- **Swipe down with two fingers on a list** → Quickly select multiple items

- **Drag multiple apps at once** → Tap and hold, then tap others before moving

Back Tap Shortcut Customization

Back Tap is a hidden gem that lets you assign shortcuts to a double- or triple-tap on the back of your iPhone.

How to Enable:

Go to: **Settings > Accessibility > Touch > Back Tap**

Choose actions for:

- **Double Tap** – Take Screenshot, Open Control Center, Mute, etc.

- **Triple Tap** – Launch an app, run a Shortcut, activate Magnifier, etc.

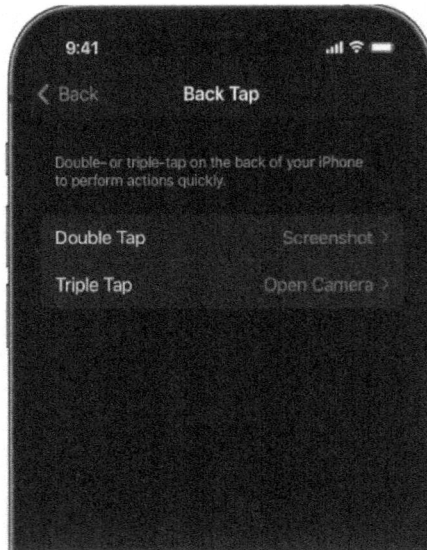

Assign smart actions to a double- or triple-tap on the back of your phone.

Advanced Focus Modes and

Automation

Focus Modes let you control what notifications you receive and when. But most users don't realize how deep this feature goes.

Create a Custom Focus:

1. Go to **Settings > Focus > + Add Focus**
2. Choose a category (Work, Fitness, Driving, etc.)
3. Customize:

   - **Allowed People & Apps**
   - **Lock Screen & Home Page**
   - **Smart Activation by time, place, or app**

Link Focus to Automations:

- Use **Shortcuts app** > Automation to trigger Focus Modes
- Examples:
  - Start "Reading Focus" when you open Kindle

- Auto-activate "Work Focus" at 9 AM on weekdays

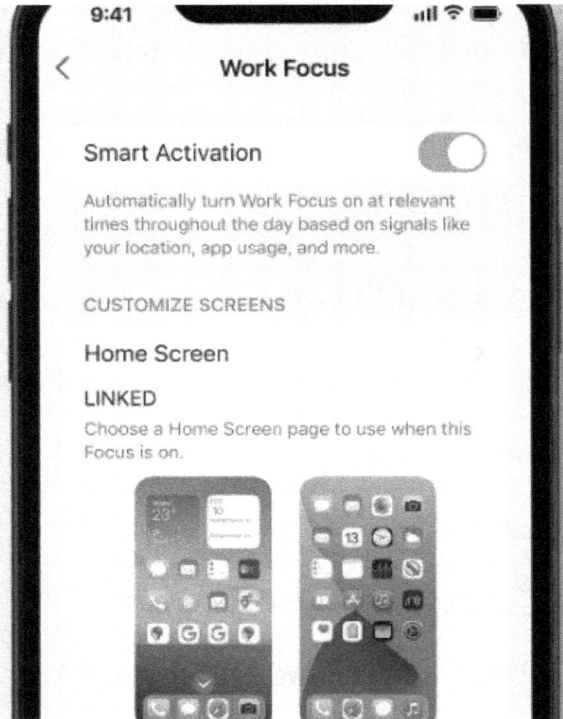

Create a distraction-free environment tailored to your lifestyle with advanced Focus settings.

How to Use Siri Shortcuts to Automate Tasks

Siri Shortcuts can automate daily routines using voice

commands, widgets, or taps. Think of them as mini-apps you design yourself.

Get Started:

Open **Shortcuts app > Gallery**

- Add pre-made shortcuts like:
  - "Text My ETA"
  - "Set Low Power Mode"
  - "Log Daily Water Intake"

Create a Custom Shortcut:

1. Tap + > Add Action
2. Choose apps like Messages, Music, or Calendar
3. Chain multiple actions into one Shortcut

Examples:

- "Good Night" shortcut: Turn on Do Not Disturb, lower brightness, play sleep music

- "I'm Home" shortcut: Text your family, adjust thermostat, open Maps

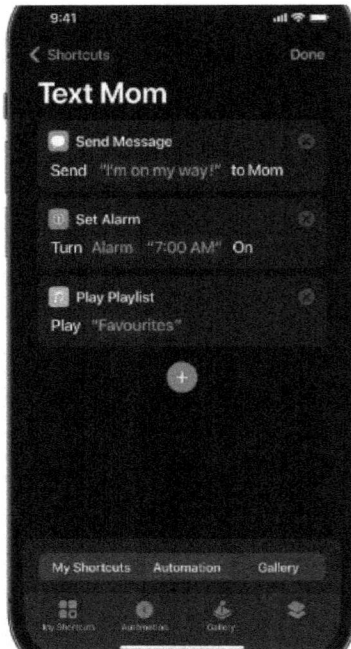

Siri Shortcuts let your iPhone do multiple things with a single tap or voice command.

Advanced Keyboard Hacks: Text Replacement, Dictation, and Voice

Control

Text Replacement

Create shortcuts for long phrases.

Go to: **Settings** > **General** > **Keyboard** > **Text Replacement**

Examples:

- "omw" → "On my way!"

- "addr" → Your full home address

- "eml" → Your email address

Dictation

Tap the **microphone icon** on the keyboard to speak your message or note aloud.

Tips:

- Say punctuation out loud (e.g., "period," "new line")

- Works in Notes, Messages, Mail, Safari, and more

Voice Control (Accessibility)

Go to: **Settings > Accessibility > Voice Control > On**

- Control your iPhone hands-free using just your voice
- Commands include "Swipe left," "Open Safari," "Tap Share"

Type faster and smarter with keyboard shortcuts, voice dictation, and custom phrases.

128

Handoff, Continuity, and Universal Clipboard with Mac/iPad

If you own multiple Apple devices, you can move seamlessly between them using Continuity features.

Universal Clipboard

- Copy on iPhone → Paste on Mac or iPad (or vice versa)
- Works with text, photos, and files

Handoff

- Start an email, note, or message on one device
- Continue exactly where you left off on another

Calls & Messages on Mac

- Make or receive iPhone calls on your Mac
- Send/receive iMessages from your Mac using the same Apple ID

Instant Hotspot

- Mac or iPad connects to iPhone's cellular hotspot automatically—no password needed

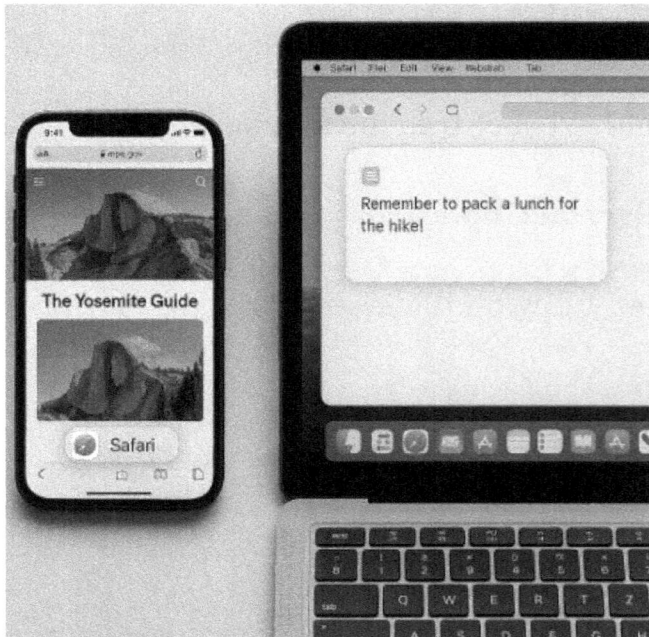

Apple Continuity keeps your work—and your life—flowing across all your devices.

Recap: Unlock the Hidden Genius of iOS 18

You now know how to:

✓ Use advanced gestures for fast, fluid navigation

✓ Customize your iPhone with Back Tap and Focus automations

✓ Build custom Siri Shortcuts to simplify your day

✓ Improve typing and accessibility with dictation and text hacks

✓ Sync work and creativity across iPhone, iPad, and Mac

✓ The tools are there. You just unlocked them.

In the next chapter, we'll take a deep dive into battery, storage, and performance optimization to keep your iPhone 16 running at its best.

Chapter 9

Battery, Storage & Performance Optimization

Your iPhone is a powerful tool, but just like any machine, it runs best when it's well maintained. This chapter shows you how to monitor battery health, manage storage, and boost speed without tech stress or guesswork.

Whether you've had your iPhone for a week or a year, these simple adjustments will help it run like new.

Understanding Battery Health & Charging Tips

Your battery isn't just about how much charge it holds—it's about how well it ages.

Check Battery Health:

Go to: **Settings > Battery > Battery Health & Charging**

You'll see:

- **Maximum Capacity** – 100% is ideal for a new device
- **Peak Performance Capability** – Confirms if your battery supports full performance

Charging Tips for Longer Battery Life:

- Don't let your battery drop to 0% often
- Avoid charging beyond 80–90% if possible
- Use **Optimized Battery Charging** (enabled by default)

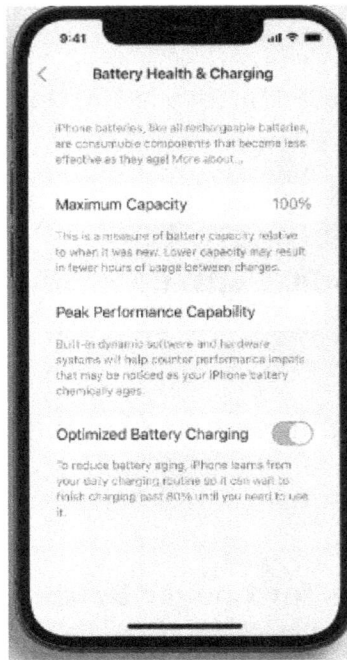

Check your battery's health to understand how it's aging—and optimize how it charges.

Optimizing Battery Usage: Low Power Mode, Background App Refresh

If your battery drains too quickly, these features can make a huge difference.

Low Power Mode

- Temporarily disables visual effects, mail fetch, and background activity

- Automatically prompts when battery drops below 20%

Turn it on anytime via:

Settings > Battery > Low Power Mode

Or use **Control Center toggle**

Background App Refresh

- Prevents apps from updating in the background
- Saves both battery and data

Go to: **Settings > General > Background App Refresh > Off or Wi-Fi Only**

Reduce unnecessary drain by controlling what runs behind the scenes.

Smart Storage: Offload Apps, Manage Photos, Clear Cache

Running low on space? You don't need to delete everything—just be strategic.

Offload Unused Apps

Go to: **Settings > iPhone Storage**

- Tap any app > Choose **Offload App**

- The icon stays, and data is preserved

Enable automatic offloading:

Settings > App Store > Offload Unused Apps

Manage Photos & Videos

- Use iCloud Photos with "Optimize iPhone Storage"

- Delete blurry, duplicate, or unused shots regularly

- Offload videos to iCloud, Google Photos, or external storage

Clear Safari & App Cache

- **Safari:** Go to **Settings > Safari > Clear History and Website Data**

- **Third-party apps:** Use in-app settings or reinstall if cache builds up

Clear clutter without losing memories by offloading apps and optimizing storage.

iPhone Clean-Up Checklist for

Speed

Here's your quick action list to boost speed and free up space:

Weekly Checklist:

- Offload apps you haven't used in 30 days

- Delete old messages with large attachments

- Remove duplicate photos or screenshots

- Restart your iPhone to refresh memory

- Close background apps (swipe up from bottom and flick away)

- Clear Safari history

Monthly Deep Clean:

- Backup photos and videos to cloud or computer

- Re-evaluate storage-heavy apps

- Check Battery Health for aging signs

- Update iOS for speed and security fixes

iPhone Clean-Up Checklist for Speed

Delete unused apps ☑

Remove old photos ☑

Clear message history ☑

Clean up your browser ☑

Optimize cloud storage ☑

Follow this checklist to keep your iPhone fast, clean, and responsive.

iOS 18 Performance Boosting Tips

Disable Motion Effects

- Reduce visual strain and save battery

- Go to: **Settings > Accessibility > Motion > Reduce Motion**

140

Siri & Search Tweaks

- Go to: **Settings > Siri & Search**
- Disable suggestions for unused apps to speed up Spotlight search

Disable Unused Widgets

- Long-press any widget > Tap **Remove Widget**
- Especially helpful for battery-draining live widgets

Close App Tabs

- In Safari: Tap the tabs icon > Close All Tabs
- Too many open tabs slow browsing and drain memory

Recap: Power, Storage, and Speed Mastery

In this chapter, you've learned how to:

- ✓ Check and care for your iPhone battery
- ✓ Use Low Power Mode and Background App Refresh wisely

- ✓ Offload unused apps and optimize photo storage
- ✓ Clean up Safari and other app data
- ✓ Boost iOS 18 performance with smart tweaks
- ✓ Follow a simple checklist to keep your device running like new

Next up, we'll explore accessibility tools that make the iPhone more usable for all—whether you have visual, physical, or hearing-related needs.

Chapter 10

Privacy & Security Settings Every User Must Know

Your iPhone knows a lot about you—your face, location, habits, passwords, and even your heart rate. Thankfully, Apple has built iOS 18 with **strong privacy** and s**ecurity tools**—you just need to know how to use them.

This chapter guides you through the essentials to keep your personal data safe, your identity secure, and your peace of mind intact.

Face ID & Passcode Management

Face ID is your first line of defense, but it works best when paired with a strong passcode.

Check or Change Your Passcode:

Go to: **Settings > Face ID & Passcode**

- Enter your current passcode

- Tap **Change Passcode**

- Choose a 6-digit, alphanumeric, or custom numeric code

Manage Face ID Uses:

- Toggle ON/OFF for **iPhone Unlock, App Store, Wallet, Passwords**, etc.

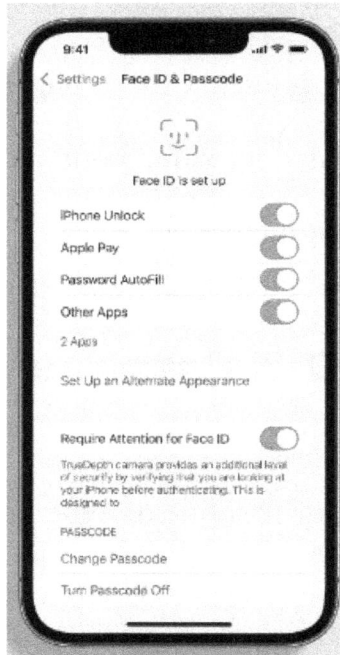

Keep your iPhone secure with a strong passcode and manage what Face ID can unlock.

iCloud Keychain, Safari Autofill, and Password Recommendations

iCloud Keychain

Securely stores your passwords, credit cards, and Wi-Fi logins across Apple devices.

Enable via:

Settings > Apple ID > iCloud > Passwords and Keychain

Safari Autofill

- Automatically fills login info on websites and apps
- Go to: **Settings > Safari > Autofill**
- Manage contact info, credit cards, and saved passwords

Password Suggestions & Security Alerts

- Go to: **Settings > Passwords**

145

- Tap any account to view or edit

- Watch for **"Security Recommendations"** if a password is reused or breached

Hide My Email & Mail Privacy Protection

Apple gives you tools to protect your real email and stop marketers from tracking your mail activity.

Hide My Email (iCloud+ feature)

- Generates a random email alias that forwards to your inbox

- Use when signing up for newsletters, apps, or websites

- Go to: **Settings > Apple ID > iCloud > Hide My Email**

Mail Privacy Protection

- Hides your IP address

- Prevents senders from knowing when you've opened an email

Enable via: **Settings > Mail > Privacy Protection**

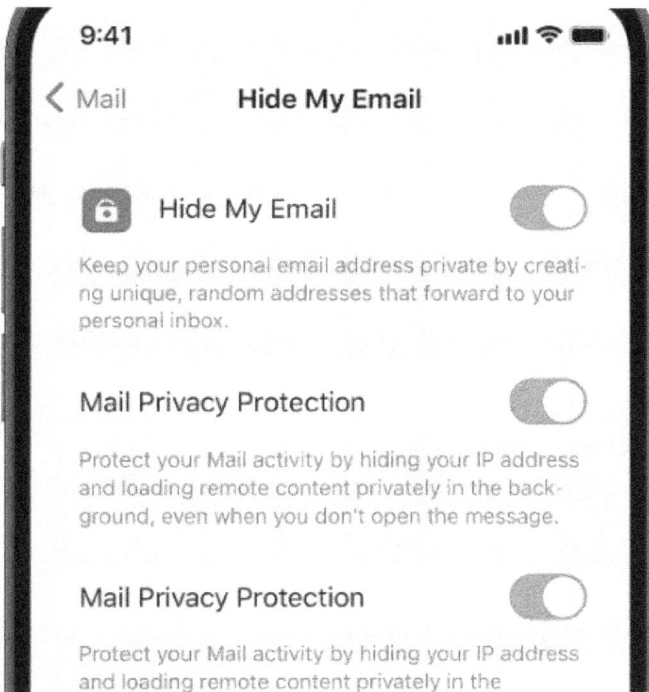

Caption: Protect your inbox identity and activity with Hide My Email and Mail Privacy Protection.

App Tracking Transparency and Permissions

iOS gives you full control over how apps collect your data.

App Tracking Transparency

When opening a new app, you'll be asked:

"**Allow [App] to track your activity across other companies' apps and websites**?"

- Tap **Ask App Not to Track** to block

Review permissions later at:

Settings > Privacy & Security > Tracking

Review App Permissions

Go to: Settings > Privacy & Security

- Check each category (Location, Camera, Microphone, Contacts, etc.)
- Revoke access from apps that don't need it

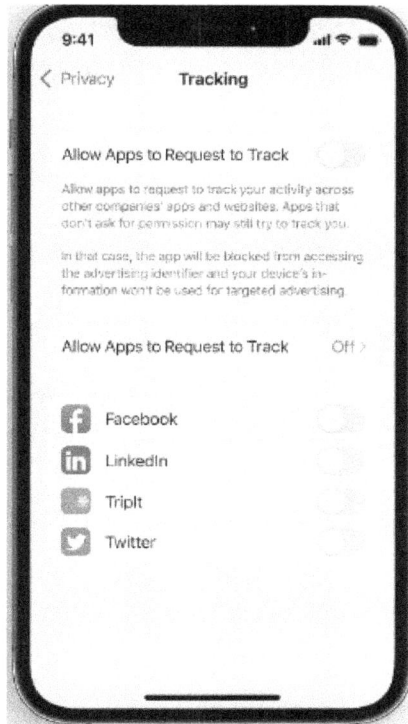

Block apps from following your digital footprint and take back control of your privacy.

Locking Notes, Hiding Photos, and Emergency SOS Settings

Lock Sensitive Notes

- Open Notes > Choose a note > Tap the **Share icon** > **Lock Note**

- Uses Face ID or passcode to protect personal entries

Hide Private Photos

- Select any photo > Tap **More (...)** > **Hide**

- Hidden photos are moved to **Albums** > **Hidden**

- Hide the Hidden Album under **Settings** > **Photos**

Emergency SOS

Press and hold the **Side button + Volume button** to activate Emergency SOS

Customize contacts and medical info:

Settings > Emergency SOS and Settings > Health > Medical ID

Screen Time & Parental Controls

Screen Time helps you monitor your own phone use—and control a child's device if needed.

Enable Screen Time:

Go to: **Settings > Screen Time > Turn On Screen Time**

- Set **App Limits, Downtime,** and **Content Restrictions**

Set Parental Controls:

- Choose **This is My iPhone** or **This is My Child's iPhone**
- Create a passcode to prevent changes
- Restrict app usage, explicit content, in-app purchases, and more

App Limits
Set time limits for apps.   ›

Communication Safety
Limit communication for configured   ›
Messages and AirDrop contacts.

Content & Privacy Restrictions   ›
Block inappropriate content

Daniel
Set up Screen Time for families

Daniel   ›

Set healthy boundaries for yourself or your family using Screen Time and parental controls.

Recap: Stay Safe, Stay Private

With a few smart settings, you now know how to:

✓ Use Face ID and strong passcodes for secure access

✓ Let iCloud manage your passwords safely

✓ Hide your real email and protect your mail activity

✓ Deny apps from tracking your activity

✓ Lock personal notes, hide photos, and set up Emergency SOS

✓ Use Screen Time and Parental Controls for safer family use

In the next chapter, we'll dive into accessibility features that make your iPhone more intuitive for all abilities.

Chapter 11

iPhone Photography for Creators & Influencers

The iPhone 16 isn't just for capturing memories—it's a mobile studio in your pocket.

Whether you're building a brand, running a business, or sharing your passion, this chapter will show you how to create scroll-stopping photos and videos for platforms like Instagram, YouTube Shorts, and TikTok—without needing fancy gear or advanced skills.

Best Settings for Instagram, YouTube Shorts, and TikTok

Instagram (Posts, Stories, Reels)

- **Photo Resolution:** Shoot in 4:3, then crop to square or 4:5 in editing

- **Video:** Use **1080p at 30fps or 4K at 24fps** for cinematic look

- **Reels:** Vertical, full-screen (1080x1920)

YouTube Shorts

- Record vertically in **1080p or 4K**

- Keep content under **60 seconds**

- Use subtitles or voice-over for clarity

TikTok

- 1080p vertical videos

- Prioritize lighting and audio clarity

- Record directly in Camera app, then upload to TikTok for better quality

Instagram   Shorts   TikTok

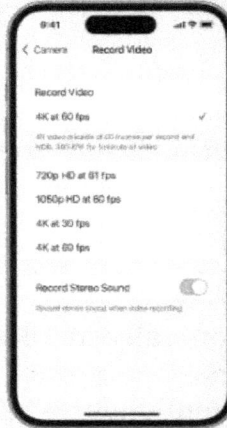

iPhone Video Settings

Tailor your iPhone's camera settings to match the requirements of each social platform.

Editing with Snapseed, Lightroom Mobile, and VSCO

These apps offer professional-grade editing tools that are easy to learn—even for beginners.

Snapseed (Free)

- Great for one-tap presets and healing tool
- Add drama, blur, vignette, and detailed tuning

Lightroom Mobile (Free with optional paid tools)

- Precision editing for exposure, contrast, color grading
- Use selective edits and presets for Instagram aesthetic

VSCO (Free/Paid)

- Stylish filters (F2, A6, HB2) with analog film vibe
- Fine-tune skin tone, grain, highlights, and shadows

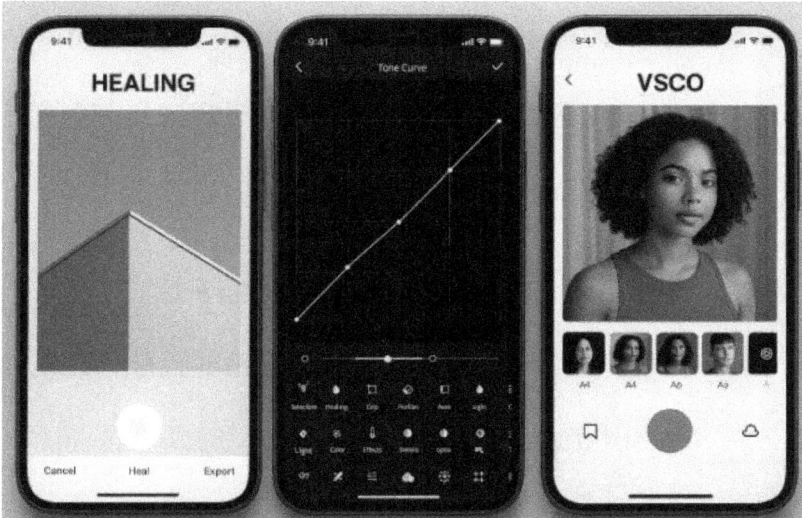

Use editing apps to add personality and polish to your iPhone photos before posting.

Using Halide and ProCamera for DSLR-like Control

Want more control over your camera settings like ISO, shutter speed, and focus? These apps unlock professional-level tools.

Halide

- Manual focus with focus peaking
- Shoot in **RAW** for ultimate editing flexibility
- Histogram, ISO, shutter, and white balance controls

ProCamera

- Shoot in **TIFF, RAW, or HEIF**
- Manual video settings: frame rate, stabilization, resolution
- Anti-shake mode and portrait depth control

Halide   ProCamera

Take full creative control with advanced manual camera apps built for professionals.

iPhone Gimbals, Tripods, and Lenses (What to Buy)

Enhance your mobile filmmaking and photography with a few creator-friendly accessories:

Top Gimbals

- **DJI Osmo Mobile 6** – 3-axis stabilization for buttery smooth video
- **Zhiyun Smooth 5** – Great for cinematic panning and motion shots

Tripods

- **Joby GorillaPod** – Flexible legs for any surface or pole
- **UBeesize iPhone Tripod** – Lightweight, adjustable height, Bluetooth shutter

Lenses

- **Moment Lenses** (Wide, Tele, Macro) – Screw-on lens kits with professional glass
- Use with **Moment or Halide apps** for best results

A few simple accessories can make your iPhone content look like it came from a pro studio.

Shooting in RAW: What It Means and How to Use It

RAW photos contain unprocessed image data, giving you full control in post-editing but they also take up more space.

How to Enable RAW:

- In Camera app (Pro models): Tap **RAW** button (top corner)

- Or go to: **Settings > Camera > Formats > Apple ProRAW ON**

When to Use RAW:

- Portraits or landscapes with strong contrast

- Product photography or food shots

- When you plan to edit heavily (e.g., Lightroom or Photoshop)

Caption: Capture every detail and shadow—shoot in RAW when quality and control matter most.

Making the Most of Cinematic Video and Audio Editing Apps

Apple's Cinematic Mode lets you create movie-style videos with shallow depth of field and subject tracking.

Cinematic Mode Tips:

- Record in **24fps or 30fps** for natural motion

- Tap to shift focus between subjects

- Edit depth and focus points after recording

Best Video Editing Apps:

- **CapCut** – Easy transitions, effects, and text overlays

- **LumaFusion** – Pro-grade timeline editing, multi-cam support

- **InShot** – Trim, cut, filter, and add music easily

Best for Audio:

- **Ferrite Recording Studio** – Voiceovers and podcasting

- **Voice Memos + iMovie** – Combine narration with video

Recap: From Camera to Creator

You now know how to:

✓ Optimize your shots for social platforms

✓ Edit like a professional using Snapseed, Lightroom, and VSCO

✓ Unlock DSLR-style control with Halide and ProCamera

✓ Use essential gear to stabilize and enhance your shots

✓ Capture and edit RAW photos

✓ Make cinematic videos and tell compelling stories through audio and editing

You don't need a studio—just your iPhone, your eye, and a little creative vision.

Chapter 12

Troubleshooting & Real-World Scenarios

Even the best technology occasionally misbehaves.

Whether your iPhone won't turn on, an app keeps crashing, or you've lost the device altogether—don't panic. Most problems can be solved with a few quick actions, and this chapter walks you through every common scenario.

iPhone Won't Charge or Turn On? Try This First

If your iPhone won't respond, it doesn't always mean it's broken.

Step 1: Check Charging Cable & Port

- Try a different charging cable and adapter

- Inspect for debris inside the Lightning or USB-C port

- Gently clean using a dry soft brush

Step 2: Force Restart

- Press and quickly release **Volume Up**, then **Volume Down**, then hold the **Side button** until the Apple logo appears.

Step 3: Try Another Power Source

- Plug into a different wall outlet or USB port

- Use a computer or power bank to test

Forgotten Passcode or Face ID Fails?

Face ID may fail after multiple unsuccessful attempts or when the phone hasn't been unlocked in a while.

If Face ID Fails:

- Swipe up and enter your passcode manually

- Clean your front camera lens

- Re-scan your face: **Settings > Face ID & Passcode > Reset Face ID**

If You Forget Your Passcode:

You'll need to erase the device and restore from backup.

Use iTunes/Finder:

1. Connect iPhone to a computer

2. Open iTunes (PC/macOS Mojave) or Finder (macOS Catalina+)

3. Enter Recovery Mode:

   - Press and quickly release **Volume Up**

   - Press and quickly release **Volume Down**

   - Hold **Side button** until Recovery Mode screen appears

4. Click **Restore**

Fixing Common App Crashes and Glitches

Apps freezing or crashing? Here's how to fix it.

Quick Fixes:

- **Force Quit** the app: Swipe up from bottom > Flick app off-screen

- **Restart** your iPhone

- **Update** the app via App Store

- **Delete & Reinstall** the app (your data may sync via iCloud)

Clear Safari Crashes:

Go to **Settings** > **Safari** > **Clear History and Website Data**

Wi-Fi, Bluetooth, and Cellular

Troubleshooting

Wi-Fi Fixes:

- Toggle Wi-Fi OFF and ON
- Forget and reconnect: **Settings > Wi-Fi > (i) > Forget This Network**
- Restart router

Bluetooth Fixes:

- Go to: **Settings > Bluetooth > Forget Device > Reconnect**
- Turn off other connected Bluetooth accessories

Cellular Data Fixes:

- Enable Airplane Mode for 10 seconds, then disable
- Go to: **Settings > General > Transfer or Reset iPhone > Reset > Reset Network Settings**

What to Do If You Lose Your iPhone

(Find My, Lost Mode, etc.)

Use Find My iPhone:

- Go to: iCloud.com/find or use **Find My app** on another Apple device

- Select your iPhone > Tap **Play Sound, Mark As Lost,** or **Erase iPhone**

Enable Lost Mode:

- Locks your iPhone

- Displays a custom message with your contact info

- Tracks its location even if offline

Important: Ensure **Find My** is enabled:

Settings > [Your Name] > Find My > Find My iPhone > On

How to Reset Settings Without

Losing Everything

If your iPhone behaves oddly but you don't want to erase your photos and apps:

Reset Settings Only:

- Go to: **Settings > General > Transfer or Reset iPhone > Reset > Reset All Settings**

This will:

- Reset Wi-Fi networks, wallpapers, Face ID, and preferences
- Leave photos, messages, and apps untouched

Other Reset Options:

- Reset Keyboard Dictionary
- Reset Home Screen Layout
- Reset Location & Privacy

When to Contact Apple Support &

How

When You Should Reach Out:

- Physical damage (screen, water, etc.)

- Repeated system crashes

- Activation or Apple ID issues

- Unresponsive Face ID or camera hardware

How to Get Help:

1. Visit: support.apple.com

2. Use the **Apple Support app**

3. Call Apple Support:

    - In the U.S.: 1-800-MY-APPLE

    - In other regions: See website for local number

Genius Bar Appointments:

- Open **Apple Support app** > **Get Support** > **Schedule a Repair**

- Or visit your nearest Apple Store

173

Recap: Stay Calm, Troubleshoot Smart

In this chapter, you learned how to:

- ✓ Fix an iPhone that won't charge or turn on

- ✓ Recover from passcode or Face ID lockouts

- ✓ Solve app crashes and connection issues

- ✓ Track and lock a lost iPhone

- ✓ Reset settings safely

- ✓ Know when and how to contact Apple Support

No more tech panic. You've got the tools to troubleshoot like a pro.

BONUS CHAPTER

Daily Practice – 30 Days to iPhone Confidence

Learning your iPhone doesn't require hours of instruction. In fact, just 5–10 minutes a day can make you a confident user in a single month.

This chapter gives you a 30-day roadmap filled with fun, stress-free tasks plus creative challenges to stretch your skills and celebrate your new digital independence.

A Day-by-Day Beginner Task List (Start Here, Grow Daily)

Each task below builds on the previous one. You'll start with the basics and grow steadily into advanced comfort.

Week 1: The Essentials

- **Day 1:** Turn your iPhone on/off, adjust volume and brightness
- **Day 2:** Make a call, save a new contact
- **Day 3:** Send a text, try emojis ☺
- **Day 4:** Take a photo and video
- **Day 5:** Connect to Wi-Fi and Bluetooth
- **Day 6:** Explore Safari and bookmark a website
- **Day 7:** Customize your Home Screen (move or delete apps)

Week 2: Daily Life with iPhone

- **Day 8:** Add a calendar event
- **Day 9:** Set an alarm and timer
- **Day 10:** Use Notes or Reminders
- **Day 11:** Try FaceTime with a friend
- **Day 12:** Use Maps to get directions
- **Day 13:** Search for and install your first app

- **Day 14:** Create your Favorites list in Contacts

Week 3: Smarter Use & Security

- **Day 15:** Set up Face ID and check Battery Health
- **Day 16:** Explore Control Center shortcuts
- **Day 17:** Use Voice Dictation in Messages or Notes
- **Day 18:** Learn how to offload unused apps
- **Day 19:** Clean up your photo gallery
- **Day 20:** Set up Find My iPhone
- **Day 21:** Clear your Safari browser cache

Week 4: Confidence & Customization

- **Day 22:** Create your own custom Focus Mode
- **Day 23:** Organize apps into folders
- **Day 24:** Send an audio message
- **Day 25:** Try a new widget on your Home Screen
- **Day 26:** Change your wallpaper and lock screen style
- **Day 27:** Back up your phone to iCloud

- **Day 28:** Test Emergency SOS or Medical ID

- **Day 29:** Teach someone else a trick you've learned

- **Day 30:** Celebrate—take a selfie and share it proudly!

Weekly Challenges for Advanced Skills

Each Sunday, challenge yourself with a deeper dive.

- **Week 1 Challenge:** Use Siri to complete 3 tasks hands-free

- **Week 2 Challenge:** Record and edit a 10-second video using cinematic mode

- **Week 3 Challenge:** Create a Shared Album and invite someone to add photos

- **Week 4 Challenge:** Set up a Shortcut to send a daily reminder or message

Weekly Challenges for Advanced Skills

S	M	T	W	T	F	S

Week 1 Challenge
Use Siri to complete 3 tasks hands-free

Week 2 Challenge
Record and edit a 10-second video

Week 3 Challenge
Create a Shared Album and invite someone

Week 4 Challenge
Set up a Shortcut to send a daily reminder

S	M	T	W	T	S

Each week, build more confidence by trying a single, deeper-level task.

Photography Prompt Challenges to Boost Your Creativity

You don't need a reason to take a great photo—just a prompt.

179

Try these over the month:

- A flower in morning light ✿
- Your favorite meal 🍜
- Something red ♥
- A reflection in a window or puddle ⬤
- Your shoes from above 👟
- A shadow pattern ☀
- A candid moment of someone laughing ☺
- A photo that tells a story in one shot 📖
- A before-and-after comparison 📷
- A memory from your day

Your First Automated Shortcut

Create your first Siri Shortcut in less than 2 minutes:

1. Open the **Shortcuts** app
2. Tap +, then **Add Action**
3. Search for "Send Message" → tap it

4. Enter recipient and text like "Good morning!"

5. Tap **Done**

6. Say: "Hey Siri, good morning!"

Now you've automated your first interaction.

Building Your "Essential Apps" Home Screen

Once you've explored more apps, rearrange your most-used into a single swipe.

Suggested Layout:

- **Top Row:** Phone, Messages, Camera, Safari

- **Row 2:** Calendar, Reminders, Maps, Weather

- **Row 3:** Photos, App Store, Mail, Notes

- **Dock:** Music/Spotify, iMessage, FaceTime, Settings

Long-press any app → Tap **Edit Home Screen** → Drag

into place or folder

Celebrating Your New iPhone Mastery

You've made it.

You've learned to text, call, search, shoot, edit, organize, troubleshoot, and even automate. This is more than mastering a phone, it's about mastering confidence in the digital age.

Now:

- Help someone else learn
- Explore more Shortcuts or Camera apps
- Or simply enjoy how seamless your iPhone feels today

You're not just a beginner anymore. You're the person people ask for help.

Acknowledgments

To every beginner, senior, and curious learner who picked up this book—thank you. Your desire to grow, adapt, and explore the digital world inspires everything I write.

Special thanks to my family and friends for their encouragement, to my team for their unwavering support, and to every readers whose feedback helped guide this book into something truly useful.

www.ingramcontent.com/pod-product-compliance
Lightning Source LLC
Chambersburg PA
CBHW031853200326
41597CB00012B/398